Nora Four

A Numberline Lane book

by

Fiona and Nick Reynolds

"What a chilly day," Nora Four said to herself.

Snow had been falling to the ground all night long.

Nora Four had just made herself a steaming cup of tea.

Outside, the ground was covered with a milky white blanket and all seemed still.

Nora Four looked out of her window. She had never seen so much snow.

In fact, she couldn't remember the last time it had snowed down Numberline Lane.

Suddenly there was a loud "Knock–knock, knock–knock, knock–knock," on her front door.

"Whoever could that be?" thought Nora Four.

Then the sound came again, "Knock–knock, knock–knock, knock–knock."

"It must be Nick Six," said Nora Four to herself.

She opened the door, and there, shivering in front of her, was Nick Six.

"Hello, Nick Six. Do come in," said Nora Four.

"Hello, Nora Four," said Nick Six. "I wonder if you could help me."

"What is the problem?" asked Nora Four as she poured him a cup of tea.

Nick Six explained that he had been trying to play his guitar, but his fingers were so cold he kept playing the wrong notes.

Nigel Nine was having a similar problem.

His fingers were so cold he couldn't make any more cakes.

Jenny Ten had such cold fingers that she could no longer hold her paintbrush and Kevin Seven was so cold that he couldn't hold his hammer.

"Hmmm! What a shame!" said Nora Four.

"We were wondering if you could knit us some gloves please?" asked Nick Six.

"Well, I'm sure I could help. Now let me think ... how many gloves will I need to make?"

Suddenly there was a flash of light and a puff of smoke.

There, standing in the middle of Nora Four's house was Gus Plus.

He looked very cosy in his woolly scarf, warm gloves, and matching hat.

"Hello, Nora Four," said Gus Plus, "Would you like some help?"

"Oh, yes please!" said Nora Four.

She explained that everyone from Numberline Lane would like some gloves, but she didn't know how many to knit.

With another flash of light and a puff of smoke, a very special numberline appeared that began "2, 4, 6, …".

"All you need to do,
 is keep on adding two!"

said Gus Plus, and, just as quickly as he had appeared, he vanished with a final puff of smoke.

Nora Four and Nick Six started adding two to build the new numberline.

"Six add two makes eight," said Nora Four.

"Eight add two makes ten," said Nick Six.

They kept going until they had reached all the way to twenty.

They looked at the numberline together and counted very carefully.

"Two, four, six, eight, ten, twelve, fourteen, sixteen, eighteen, twenty."

"If I made one pair, that would be two gloves. If I made two pairs, that would be four gloves."

"If I made three pairs, that would be six gloves. I will make ten pairs so everyone can have some. That will be twenty gloves to knit."

So Nora Four got to work. She chose lots of brightly coloured wool and soon had knitted everyone a pair of gloves.

She put on her own pair and went outside into the snow.

There she found everyone from Numberline Lane with Gus Plus.

She handed out the gloves and they all said thank you.

Linus Minus, who had been hiding behind a tree, threw a snowball at Gus Plus.

Gus Plus turned around and saw Linus Minus chuckling to himself.

Gus Plus picked up some snow and threw it at Linus Minus.

Soon all the Numbers were enjoying throwing snowballs.

None of them had cold fingers – except Linus Minus.

Nora Four said that she would make him some of his very own gloves as soon as she got home.

And she did.